THE LADDER OF MATHEMATICS
Understanding Mathematics Through Theorems and Definitions

By Wisdom Kwaku Ameyedowo, MPhil

THE LADDER OF MATHEMATICS

Understanding Mathematics

Through Theorems and Definitions

By Wisdom Kwaku Ameyedowo, MPhil

Introduction

First of all, I love Theorems and Definitions. Because these portray how Mathematicians think and reason.

I consider myself to be a Student of Mathematics. And this book is about making Mathematics more accessible to the understanding of everyone eager enough to venture into the world of Mathematics.

So this book is actually for everyone. The Theorems and Definitions are chosen to instill a certain level of curiosity, questioning and interest into the subject of Mathematics. It also seeks to present Mathematics as a language.

So I have selected these Theorems and Definitions from various sources spanning the vast spectrum of Mathematics. This is based on my long years of study and research into the subject of Mathematics into finding the best way to understand it without much effort.

So that the central idea here is to get familiarity with its laws and let them sink into the mind as seeds. These seeds, nurtured and cherished, will eventually sprout into beautiful structures that we usually find in Mathematics that put many away at first sight.

So this book is for students, lecturers, professors, researchers in Mathematics, and anyone that wants to know the Language of Mathematicians in an easy way.

Wisdom Kwaku Ameyedowo, MPhil

Accra, November 2015

CHAPTER 1

1. *If M is a module over a commutative ring R and $a \in M$, then $\mathcal{O}_a = \{r \in R; ra = 0\}$ is an ideal of R. If $\mathcal{O}_a \neq 0$, a is said to be a **torsion element** of M.*

2. *Let $f: X \to Y$ be a map between topological spaces. Suppose that every point $p \in X$ has a neighborhood U on which the restriction $f|_U$ is continuous, then f is continuous.*

3. *An algebraic extension field F of K is **normal** over K (or a **normal extension**) if every irreducible polynomial in $K[x]$ that has a root in F actually splits in $F[x]$.*

4. ***Rank − Nullity law***
   Let $T: V \to W$ be a linear map between finite − dimensional vector spaces. Then $\dim V = rank\ T + nullity\ T = \dim(Im\ T) + \dim(\ker T)$.

5. *Let X be a topological space and $x \in X$ a point. On the set $\mathcal{P}(X)$ of all subsets of X we define an equivalence relation: $A \sim_x B$ iff $A \cap U = B \cap U$ for some neighbourhood U of x. An equivalence class (A, x) of the relation $\sim_x$ is called a **germ of a subset of** X **at the point** x.*

6. ***Gronwall's Inequality***
   Let $r, k \in C([a,b], \mathbb{R})$ and suppose that $r(t) \geq 0$ and $k(t) \geq 0$ for all $t \in [a,b]$. Let δ be a given nonnegative constant. if
   $$r(t) \leq \delta + \int_a^t k(s)\, r(s)\, ds \text{ for all } t \in [a,b], \text{then}$$

3

$r(t) \leq \delta e^{\int_a^t k(s)ds}$ *for all* $t \in [a, b]$.

7. *Any closed and bounded nonempty set of real numbers has a minimum and a maximum.*
8. *Let* $\mathbb{Z}[i]$ *be the following subset of the complex numbers* $\mathbb{Z}[i] = \{a + bi; a, b \in \mathbb{Z}\}$. *Then* $\mathbb{Z}[i]$ *is an integral domain called the domain of* **Gaussian integers**.

9. *Let* X *be an ordered set in the order topology and* y *an interval or a ray in* X, *then the subspace topology and the order topology on* Y *are the same.*

10. *A map* $f: X \to Y$ *is said to be an* **open map** *if for every open set* U *of* X, *the set* $f(U)$ *is open in* Y.

11. *Let* $0 \to L \overset{\alpha}{\to} M \overset{\beta}{\to} N \to 0$ *be a short exact sequence of* R − *modules. Then the following three conditions are equivalent:*
   (i) there exists an isomorphism $M \cong L \oplus N$ *under which* α *is given by* $m \mapsto (m, 0)$ *and* β *by* $(m, n) \mapsto n$;
   (ii) there exists a **section** *of* β, *that is, a map* $s: N \to M$ *such that* $\beta \circ s = Id_N$;
   (iii) there exists a **retraction** *of* α, *that is, a map* $r: M \to L$ *such that* $r \circ \alpha = Id_L$.

If this happens the sequence is a **split exact sequence**.

12. *A morphism* $f: X \to Y$ *of schemes of finite type over* k *is* é**tale** *if it is smooth of relative dimension* 0.

13. *Let* $\Omega \subset \mathbb{C}^n$ *be a connected open subset and* $A \subset \Omega$ *a complex analytic subset. Then* $\Omega \backslash A$ *is dense and connected.*

14. *Let* G *be a group. The subgroup of* G *generated by the set*

$\{aba^{-1}b^{-1}; a, b \in G\}$ is called the **commutator subgroup** of G.

15. The nonempty connected subsets of the real line are the intervals.

16. Chow's Lemma

Let X be proper over a noetherian scheme S. Then there is a scheme X' and a morphism $g: X' \to X$ such that X' is projective over S, and there is an open dense subset $U \subseteq X$ such that g induces an isomorphism of $g^{-1}(U)$ to U.

17. If V is a finite − dimensional inner product space and $S \subset V$ is a subspace, the **orthogonal complement** of S in V is the set $S^{\perp} = \{x \in V; \langle x, y \rangle = 0 \text{ for all } y \in S\}$.

18. Inverse Functions Theorem

If the derivative $d_x F: \mathbb{R}^k \to \mathbb{R}^k$ is nonsingular, then F maps any sufficiently small open set U about x diffeomorphically onto an open set $F(U)$.

19. The point $a \in A$ is called an **isolated point** of A if it has a neighborhood whose intersection with A reduces to the point a.

20. Let $f: M \to N$ be a proper map between topological spaces, and $A \subset M$ any subset that is saturated with respect to F. Then $f|_A: A \to f(A)$ is proper.

21. An algebraic set $X \subset \mathbb{A}_k^n$ is **irreducible** if there does not exist a decomposition $X = X_1 \cup X_2$ with $X_1, X_2 \subsetneq X$ of X as a union of two strictly algebraic subsets.

22. Noether Normalisation Lemma

Let k be an infinite field, and $A = k[a_1, ..., a_n]$ a finitely generated k − algebra. Then there exists $m \leq n$ and $y_1, ..., y_m \in A$ such that (i) $y_1, ..., y_m$ are algebraically independent over k; and

(ii) A is a finite $k[y_1, \ldots, y_m]$ − algebra.

23. *A scheme is **normal** if all of its local rings are integrally closed domains.*

24. *Every vector space V over a division ring D has a basis and is therefore a free D − module. More generally, every linearly independent subset of V is contained in a basis of V.*

25. *Let X and Y be topological spaces. A continuous injective map $f: X \to Y$ is called a **topological embedding** if it is a homeomorphism onto its image $f(X) \subset Y$ in the subspace topology.*

26. *Suppose M is a compact space and N is a Hausdorff space. Then every continuous map $F: M \to N$ is proper.*

27. *A **ring** $(R, +, \cdot)$ is a set R together with two binary operations $+: R \times R \to R$ (addition) and $\cdot: R \times R \to R$ (multiplication) satisfying the following properties:*
(i) $(R, +)$is an abelian group. We write the identity element as 0.
(ii) $a \cdot (b \cdot c) = (a \cdot b) \cdot c$ ($\cdot$ is associative).
(iii) $a \cdot (b + c) = a \cdot b + a \cdot c$ and $(b + c) \cdot a = b \cdot a + c \cdot a$
($\cdot$ is left and right distributive over $+$).

28. *Let A be a subset of the topological space X; let A' be the set of all limit points of A. Then $\bar{A} = A \cup A'$.*

29. *Let X be a topological space with topology $\mathfrak{J}$. If Y is a subset of X, the collection $\mathfrak{J}_Y = \{Y \cap U; U \in \mathfrak{J}\}$is a topology on Y, called the **subspace topology**.*

30. *The integral closure of a Dedekind domain in a finite extension field of its quotient field is again a Dedekind domain.*

CHAPTER 2

1. **Morera's Theorem**

   If $f(z)$ is defined and continuous in a region Ω, and if $\int_{\Omega} f \, dz = 0$
   for all closed curves γ in Ω, then $f(z)$ is analytic in Ω.

2. Let A, B, C be modules over a commutative ring R. A **bilinear map**
   from $A \times B$ to C is a function $f: A \times B \to C$ such that for all $a, a_i \in A$,
   $b, b_i \in B$, and $r \in R$:
   (i) $f(a_1 + a_2, b) = f(a_1, b) + f(a_2, b)$;
   (ii) $f(a, b_1 + b_2) = f(a, b_1) + f(a, b_2)$;
   (iii) $f(ra, b = rf(a, b) = f(a, rb)$.

3. **First Isomorphism Theorem for Rings**
   Let $f: R \to S$ be a ring homomorphism. Then $R/Ker(f) \cong Im(f)$.

4. Let M be a topological $n-$ manifold. A **coordinate chart** (or just
   a **chart**) on M is a pair (U, φ), where U is an open subset of M
   and $\varphi: U \to \tilde{U}$ is a homeomorphism from U to an open subset
   $\tilde{U} = \varphi(U) \subset \mathbb{R}^n$.

5. **Schreier's Theorem**
   Two subnormal (normal) series of a group G have isomorphic
   refinements.

6. Let k be a fixed algebraically closed field. We define **affine**
   $n-$ **space** over k, denoted $\mathbb{A}_k^n$ or simply $\mathbb{A}^n$, to be the set of all
   $n-$ tuples of elements of k. An element $P \in \mathbb{A}^n$ will be called a
   point, and if $P = (a_1, \dots, a_n)$ with $a_i \in k$, then the a_i will be called
   the **coordinates** of P.

7. If G is a smooth manifold with a group structure such that the

map $G \times G \to G$ given by $(g, h) \mapsto gh^{-1}$ is smooth, then G is a Lie group.

8. *A ring R is **subdirectly irreducible** if the intersection of all nonzero ideals of R is nonzero.*

9. *If $char(K) = p$, the map $\sigma: x \mapsto x^p$ is an isomorphism of K onto one of its subfields K^p.*

10. *An element a of a ring R is **regular** (in the sense of Von Neumann) if there exists $x \in R$ such that $axa = a$. If every element of R is regular, then R is said to be a **regular ring**.*

11. *Let R be an integral domain and let m be a free $R -$ module. Then M is torsion $-$ free.*

12. *Suppose $\tilde{X}$ and X are topological spaces. A map $\pi: \tilde{X} \to X$ is called a **covering map** if $\tilde{X}$ is path connected and locally path connected, π is surjective and continuous, and each point $p \in X$ has a neighborhood U that is **evenly covered** by π, meaning that U is connected and each component of $\pi^{-1}(U)$ is mapped homeomorphically onto U by π.*

13. *Let R be a ring with identity and let $I \neq R$ be an ideal of R. Then there is a maximal ideal of R containing I.*

14. *Let K be a field. The image of $\mathbb{Z}$ in K is an integral domain, hence isomorphic to $\mathbb{Z}$ or $\mathbb{Z}/p\mathbb{Z}$, where p is prime; its field of fractions is isomorphic to $\mathbb{Q}$ or to $\mathbb{Z}/p\mathbb{Z} = \mathbb{F}_p$. In the first case, one says that K is of **characteristic zero**; in the second case, that K is of **characteristics** p. The characteristic of K is denoted by $char(K)$. If $char(K) = p \neq 0$, p is also the smallest integer $n > 0$ such that $n \cdot 1 = 0$.*

15. Let H be a finite group of order n. Suppose that, for all divisors d of n, the set of $x \in H$ such that $x^d = 1$ has at most d elements. Then H is cyclic.

16. Suppose that $f(z)$ is analytic at z_0, $f(z_0) = w_0$, and that $f(z) - w_0$ has a zero of order n at z_0. If $\epsilon > 0$ is sufficiently small, there exists a corresponding $\delta > 0$ such that for all a with $|a - w_0| < \delta$ the $f(z) = a$ has exactly n roots in the disk $|z - z_0| < \epsilon$.

17. Let V be a module over a commutative ring A. A function $Q: V \to A$ is called a **quadratic form on V** if:
(i) $Q(ax) = a^2 Q(x)$ for $a \in A$ and $x \in V$
(ii) The function $(x, y) \mapsto Q(x + y) - Q(x) - Q(y)$ is a bilinear form.

18. Let X be a scheme. The kernel, cokernel, and image of any morphism of quasi $-$ coherent sheaves are quasi $-$ coherent. Any extension of quasi $-$ coherent sheaves is quasi $-$ coherent. If X is noetherian, the same is true for coherent sheaves.

19. Let m be an integer ≥ 1. We denote by $G(m)$ the multiplicative group $(\mathbb{Z}/m\mathbb{Z})^*$ of invertible elements of the ring $\mathbb{Z}/m\mathbb{Z}$. It is an abelian group of order $\phi(m)$, where $\phi(m)$ is the Euler $\phi -$ function of m. An element χ of the dual of $G(m)$ is called a **character modulo m**; it can be viewed as a function defined on the set of integers prime to m, with values in $\mathbb{C}^*$, and such that $\chi(ab) = \chi(a)\chi(b)$; it is convenient to extend such a function to all of $\mathbb{Z}$ by putting $\chi(a) = 0$ if a is not prime to m.

20. Let M be a topological manifold with boundary.
(i) M is locally path connected.
(ii) M has at most countably many components, each of which is a connected topological manifold with boundary.
(iii) The fundamental group of M is countable.

21. An $R-$ module M is said to be **irreducible** if $\langle 0 \rangle$ and M are the only submodules of M.

22. If $(\mathcal{O}, \mathfrak{m})$ is a noetherian local domain with quotient field K, and if L is a finitely generated field extension of K, then there exists a discrete valuation ring R of L dominating $\mathcal{O}$.

23. A **subbasis** $\mathfrak{S}$ for a topology on X is a collection of subsets of X whose union equals X. The topology generated by the subbasis $\mathfrak{S}$ is defined to be the collection $\mathfrak{J}$ of all unions of finite intersections of elements of $\mathfrak{S}$.

24. Let H be a subgroup of G. Then every character of H extends to a character of G.

25. Let Y be a scheme. A (**geometric**) **vector bundle** of rank n over Y is a scheme X and a morphism $f: X \rightarrow Y$, together with additional data consisting of an open covering $\{U_i\}$ of Y, and isomorphisms $\psi_i: f^{-1}(U_i) \rightarrow \mathbb{A}^n_{U_i}$, such that for any i, j, and for any open affine subset $V = Spec\ A \subseteq U_i \cap U_j$, the automorphism $\psi = \psi_j \circ \psi_i^{-1}$ of $\mathbb{A}^n_Y = Spec\ A[x_1, \dots, x_n]$ is given by a linear automorphism θ of $A[x_1, \dots, x_n]$, i.e., $\theta(a) = a$ for any $a \in A$, and $\theta(x_i) = \sum a_{ij} x_j$ for suitable $a_{ij} \in A$.

26. In a ring with identity there are always maximal ideals.

27. An $R-$ module P is **projective** if there exists an $R-$ module P' such that $P \oplus P'$ is a free $R-$ module.

28. Two quadratic forms over k are equivalent if and only if they have the same rank, same discriminant, and same invariant ϵ.

CHAPTER 3

1. *On dit qu'un espace topologique X est* **noethérien** *si l'ensemble des ouverts de X vérifie la condition* **maximale**, *ou, ce qui revient au même, si l'ensemble des fermés de X vérifie la condition* **minimale**. *On dit que X est* **localement noethérien** *si tout $x \in X$ admet un voisinage qui est un sous − epace noethérien.*

2. *Pour qu'un anneau A ait au moin un idéal premier, il faut et il suffit que $A \neq \{0\}$.*

3. *Un anneau* **local** *est un anneau A dans lequel il existe un seul idéal maximal, qui est alors le complementaire des éléments inversibles et contient tous les idéaux $\neq A$.*

4. ***Uniformization Theorem***
   Any simply connected Riemann surface is conformally isomorphic either
   (a) to the plane $\mathbb{C}$ consisting of all complex numbers $z = x + iy$,
   (b) to the open unit disk $D \subset \mathbb{C}$ consisting of all z with $|z|^2 = x^2 + y^2$, or
   (c) to the Riemann sphere $\hat{\mathbb{C}}$ consisting of $\mathbb{C}$ together with a point at infinity.

5. *Si A et B sont deux anneaux locaux, $\mathfrak{m}$ et $\mathfrak{n}$ leurs idéaux maximaux respectifs, on dit qu'un homomorphisme $\varphi: A \to B$ est* **locale** *si $\varphi(\mathfrak{m}) \subset \mathfrak{n}$ (ou, ce qui revient au même, si $\varphi^{-1}(\mathfrak{n}) = \mathfrak{m}$).*

6. *Tout sous − espace irréductible d'un espace topologique X est contenu dans un sous − espace irréductible maximale, qui est nécessairement fermé.*

7. On appelle **espace de Kolmogoroff** un espace topologique X vérifiant l'axiome de séparation:
(T_0) Si $x \neq y$ sont deux points quelconques de X, il existe un ensemble ouvert contenant l'un des points x, y et non l'autre.

8. Every immediate attractive basin is either simply connected or infinitely connected.

9. A morphism $f: X \rightarrow Y$ of schemes is **affine** if there is an open affine cover $\{V_i\}$ of Y such that $f^{-1}(V_i)$ is affine for each i.

10. Let R be an integral domain. If P is a projective $R-$module, then P is torsion$-$free.

11. On dit qu'un espace topologique X est **irréductible** s'il est non vide et s'il n'est pas réunion de deux sous$-$espaces fermés distincts de X.

12. **Thérème de Krull**
Soient A un anneau noethérien, J un idéal de A, M un $A-$module de type fini, M' un sous$-$module de M; alors la topologie induite sur M' par la topologie $J-$préadique de M est identique à la topologie $J-$préadique de M'.

13. Let R be ring and let $I \subseteq R$. We say that I is an **ideal** of R if and only if
(i) I is an additive subgroup of R,
(ii) $rI \subseteq I$ for all $r \in R$, and
(iii) $Ir \subseteq I$ for all $r \in R$.

14. **Theorem of Böttcher**
Suppose that $f(z) = a_n z^n + a_{n+1} z^{n+1} + \cdots$, where $n \geq 2, a_n \neq 0$.
Then there exists a local holomorphic change of coordinate

$w = \phi(z)$ which conjugates f to the $n - th$ power map $w \longmapsto w^n$ throughtout some neighborhood of $\phi(0) = 0$. Furthermore, ϕ is unique up to multiplication by an $(n - 1) - st$ root of unity.

15. Soit $f: X \rightarrow Y$ un morphisme de préschémas. On dit que f est une **immersion locale en un point** $x \in X$ s'il existe un voisinage ouvert U de x dans X et un voisinage ouvert V de $f(x)$ dans Y tels que la restriction de f au préschéma induit U soit une immersion fermée de U dans le préschéma induit V. On dit que f est une immersion locale si f est une immersion locale en tout point de X.

16. **Theorem of F. and M. Riesz**
    Suppose that $f: D \rightarrow \mathbb{C}$ is bounded and holomorphic on the open unit disk. if the radial limit $\lim_{r \to 1} f(re^{i\theta})$ exists and is equal to zero for θ belonging to a set $E \subset [0, 2\pi]$ of positive Lebesgue measure, then f must be identically zero.

17. We say a scheme X is **regular in codimension one** (or sometimes **nonsingular in codimension one**) if every local ring $\mathcal{O}_x$ of X of dimension one is regular.

18. If the curvature and torsion of an affinely connected manifold M are both zero, then M is locally flat.

19. A **diffeomorphism** between smooth manifolds M and N is a smooth map $F: M \rightarrow N$ that has a smooth inverse.

20. Any localization of a regular local ring at a prime ideal is again a regular local ring.

21. A topological manifold M together with an equivalence class of differentiable structures on M is called a **differentiable**

manifold.

22. *Si A est un anneau, S une partie multiplicative de A, le morphisme canonique $Spec(S^{-1}A) \rightarrow Spec(A)$ est radiciel.*

23. *Soit X un préschéma; pour tout préschéma T, on note encore X(T) l'ensemble Hom(T, X) des morphismes $T \rightarrow X$, et les éléments de cet ensemble sont aussi appelés **points de X à valeurs dans** T.*

24. *Let F be an algebraic extension of K such that every polynomial in K[x] has a root in F. Then F is an algebraic closure of K.*

25. *Étant donnés trois ensembles P, Q, R et deus applications $\varphi: P \rightarrow R, \psi: Q \rightarrow R$, on appelle **produit fibré de P et Q au − dessus de** R (relatif à φ et ψ) la partie de l'ensemble $P \times Q$ formée des couples (p, q) tels que $\varphi(p) = \psi(q)$; on le note $P \times_R Q$.*

26. *Let $f: X \rightarrow Y$ be a proper, flat morphism of varieties over k. Suppose for some point $y \in Y$ that the fibre X_y is smooth over k(y). Then there is an open neighborhood U of y in Y such that $f: f^{-1}(U) \rightarrow U$ is smooth.*

27. *Let Y be a topological space. A function $\varphi: Y \rightarrow \mathbb{Z}$ is **upper semicontinuous** if for each $y \in Y$, there is an open neighborhood U of y, such that for all $y' \in U, \varphi(y') \leq \varphi(y)$.*

28. *Let $f: \mathbb{C} \rightarrow \mathbb{C}$ be a rational map of degree two or more. Then f has at most a finite number of cycles which are attracting or neutral.*

29. *Let E be an intermediate field of the extension $K \subset F$. A $K − automorphism \tau \in Aut_K(E)$ is said to be **extendible** to F if there exists $\sigma \in Aut_K(F)$ such that $\sigma|_E = \tau$.*

30. *Pour que $y \in Y$ soit le point générique d'une composante irréductible de Y, il faut et il suffit que le seul idéal premier de l'anneau local $\mathcal{O}_y$ soit son idéal maximal (autrement dit, que $\mathcal{O}_y$ soit de dimension zéro).*

31. *A **morphism** $\varphi: X \to Y$ between abstract nonsingular curves or varieties is a continuous mapping such that for every open set $V \subseteq Y$, and every regular function $f: V \to k, f \circ \varphi$ is a regular function on $\varphi^{-1}(V)$.*

32. *A region is **simply connected** if its complement with respect to the extended plane is connected.*

33. *Soit $\{U_\alpha\}$ un recouvrement (dont l'ensemble d'indices est non vide) d'un espace topologique X, formés d'ouverts non vides; pour que X soit irréductible, il faut et if suffit que U_α soit irréductible pour tout α, et que $U_\alpha \cap U_\beta \neq \emptyset$ quels que soient α, β.*

34. *A curve X in $\mathbb{P}^n$ is **strange** if there is a point A which lies on all the tangent lines of X.*

35. *Let F be a fielld. then $F[X]$ is a principal ideal domain (PID).*

36. *Let R be a commutative ring and let $\emptyset \neq S \subseteq R$ be a multiplicatively closed subset of R containing no zero divisors. The **localization of R away from** S is a commutative ring R_S with identity, and an injective ring homomorphism $\phi: R \to R_S$ such that for all $a \in R_S$ there are $b \in R$ and $c \in S$ such that $\phi(c)$ is a unit in R_S and $a = \phi(b)\phi(c)^{-1}$.*

CHAPTER 4

1. *Soient X un préschéma dont l'espace sous — jacent est localement noethérien, x un point de X. Les composantes irréductibles de $Spec(\mathcal{O}_x)$ sont les traces sur $Spec(\mathcal{O}_x)$ des composantes irréductibles de X contenant x. Pour qu'un ouvert $U \subset X$ soit tel que $U \cap Spec(\mathcal{O}_x)$ soit dense dans $Spec(\mathcal{O}_x)$, il faut et il suffit qu'il rencontre les composantes irréductibles de X contenant x (ce qui a lieu en particulier si U est dense dans X).*

2. *Let R be a ring with identity such that for every free R — module F, any two bases of F have the same cardinality. Then R is said to have the **invariant dimension property** and the cardinal number of any basis of F is called the **dimension** (or **rank**) of F over R.*

3. *A function $f(z)$ is analytic on an arbitrary point set A if it is analytic in some region which contains A.*

4. *Let R and R be rings and $\varphi: R \to S$ a ring homomorphism. Then every S — module A can be made into an R — module by defining $rx, (x \in A)$, to be $\varphi(r)x$. One says that the R — module structure of A is given by **pullback along** φ.*

5. *Let k be an algebraically closed field. A **curve** over k is an integral separated scheme X of finite type over k, of dimension one. If X is proper over k, we say that X is **complete**. If all the local rings of X are regular local rings, we say that X is **nonsingular**.*

6. *Let $\{a_n\}_{n=1}^{\infty}$ be a bounded sequence of real numbers. Let $\alpha = \lim_{n \to \infty} \inf a_n, \beta = \lim_{n \to \infty} \sup a_n$. Then, for every $\epsilon > 0$, there are at most a finite number of terms of the sequence which are*

outside the interval $(\alpha - \epsilon, \beta + \epsilon)$.

7. *Let* $f: D(f) \subseteq \mathbb{R} \to \mathbb{R}$ *be a map with domain* $D(f)$ *in* $\mathbb{R}$. *The map* *f is said to be* **continuous at** $x_0 \in D(f)$ *if for any open set* *V in* $\mathbb{R}$ *containing* $f(x_0)$, *there exists an open set U in* $D(f)$ *containing* x_0 *such that* $f(U) \subseteq V$.

8. *If f is uniformly continuous on a bounded set, then f is bounded.*

9. *A set* $S \subseteq \mathbb{R}$ *is said to be* **compact** *if and only if it is closed and bounded.*

10. *A set S is compact if and only if every sequence in S has a subsequence that converges to a point of S.*

11. *Let* $D \subseteq X \subseteq \mathbb{R}$ *and let* $f: D \to \mathbb{R}$ *be a function. we say that* $F: X \to \mathbb{R}$ *is an* **extension of** *f if* $F(x) = f(x) \; \forall \; x \in D$.

12. *Let* $f: (a, b) \to \mathbb{R}$ *be continuous. Then, f is uniformly continuous if and only if there exists an extension* $F: [a, b] \to \mathbb{R}$ *of f which is continuous.*

13. *An infinite integral* $\int_a^\infty a(t)dt$, *is said to* **converge** *if*

$$I(t) = \int_a^N a(t)dt \text{ tends to a finite limit as } N \to \infty, otherwise$$

the integral is said to **diverge**.

14. *Every closed subset of a compact space is compact.*

15. *Étant donné un* $A -$ *module M, on appelle* **support** *de M et on note* $Supp(M)$ *l'ensemble des idéaux premiers P de A tels que*

$M_P \neq 0.$

16. *A scheme is integral if and only if it is both reduced and irreducible.*

17. *A scheme X is **locally noetherian** if it can be covered by open affine subsets $Spec(A_i)$, where each A_i is a noetherian ring.*

18. *Soit M un A − module. Pour que $M = 0$, il faut et il suffit que $Supp(M) = \emptyset$.*

19. *Let S be a fixed scheme. A **scheme over** S is a scheme X, together with a morphism $X \to S$.*

20. *An analytic function in a region Ω whose derivative vanishes identically must reduce to a constant. The same is true if either the real part, the imaginary part, the modulus, or the argument is constant.*

21. **Correspondence Theorem**
    Let R be a ring, $I \subseteq R$ an ideal of R, and $\pi: R \to R/I$ the natural map. Then the function $S \longmapsto S/I$ defines a one − to − one correspondence between the set of all subrings of R containing I and the set of all subrings of R/I. Under this correspondence, ideals of R containing I correspond to ideals of R/I.

22. *If R is a ring with identity then the ring of **formal power series** $R[[X]]$ is defined similarly to the ring of polynomials. Specifically, $R[[X]]$ is the set of all functions $f: \mathbb{Z}^+ \to R$ with the same formulas for addition and multiplication as in $R[X]$. The only difference is that we do not assume that $f(n) = 0$ for all but a finite number of n. We generally write a formal power*

series as an expression $f(X) = \sum_{n=0}^{\infty} a_n X^n$.

23. Dans un anneau préadmissible A, un idéal premier ouvert contient tous les idéaux de définition.

24. If M is an R − module then a short exact sequence $0 \rightarrow K \rightarrow F \rightarrow M \rightarrow 0$, where F is a free R − module is called a **free presentation** of M.

25. Suppose X and Y are topological spaces, and $F: X \rightarrow Y$ is a continuous map that is either open or closed.
   (a) If F is surjective, it is a quotient map.
   (b) If F is injective, it is a topological embedding.
   (c) If F is bijective, it is a homeomorphism.

26. Let M be a free R − module. then the **free rank** of M, denoted $free - rank_R(M)$, is the minimal cardinality of a basis of M.

27. The arithmetic genus of a nonsingular projective surface is a birational invariant.

28. **Liouville's Theorem**
   A function which is analytic and bounded in the whole plane must reduce to a constant.

29. If X is a topological space, Y is a set, and $\pi: X \rightarrow Y$ is any surjective map, the **quotient topology** on Y determined by π is defined by declaring a subset $U \subset Y$ to be open if and only if $\pi^{-1}(U)$ is open in X. If X and Y are topological spaces, a map $\pi: X \rightarrow Y$ is called a **quotient map** if it is surjective and continuous and Y has the quotient topology determined by π.

30. *Suppose G is a Lie group with Lie algebra* $\mathfrak{g}$. *If* $\mathfrak{h}$ *is any Lie subalgebra of* $\mathfrak{g}$, *then there is a unique connected Lie subgroup of G whose Lie algebra is* $\mathfrak{h}$.

31. *If x is a point of a scheme X, we define an **étale neighborhood** of x to be an étale morphism* $f: U \to X$, *together with a point* $x' \in U$ *such that* $f(x') = x$.

32. *Let R be a PID and let* $p \in R$ *be a prime. Then submodules, quotient modules, and direct sums of* $p -$ *primary modules are* $p -$ *primary.*

33. ***The Maximum Principle***
If $f(z)$ *is analytic and nonconstant in a region* Ω, *then its absolute value* $|f(z)|$ *has no maximum in* Ω.

34. *If* $f(z)$ *is analytic and* $\neq 0$ *in a simply connected region* Ω, *a single* $-$ *valued analytic branch of* $\log f(z)$ *can be defined in* Ω.

35. *Let X be a topological space. A presheaf* $\mathfrak{F}$ *of abelian groups on X consists of the data:*
(a) for every open subset $U \subseteq X$, *an abelian group* $\mathfrak{F}(U)$, *and*
(b) for every inclusion $V \subseteq U$ *of open subsets of X, a morphism of abelian groups* $\rho_{UV}: \mathfrak{F}(U) \to \mathfrak{F}(V)$,
subject to the conditions:
(0) $\mathfrak{F}(\emptyset) = 0$, *where* $\emptyset$ *is the empty set,*
(1) ρ_{UU} *is the identity map* $\mathfrak{F}U) \to \mathfrak{F}(U)$, *and*
(2) if $W \subseteq V \subseteq U$ *are three open subsets, then* $\rho_{UW} = \rho_{VW} \circ \rho_{UV}$.

If $\mathfrak{F}$ *is a presheaf on X, we refer to* $\mathfrak{F}(U)$ *as the **sections** of the presheaf* $\mathfrak{F}$ *over the open set U, and we sometimes use the notation* $\Gamma(U, \mathfrak{F})$ *to denote the group* $\mathfrak{F}(U)$. *We call the maps*

ρ_{UV} **restriction** maps, and we sometimes write $s|_V$ instead of $\rho_{UV}(s)$, if $s \in \mathfrak{F}(U)$.

36. Every Cremer point is a $non-isolated$ point in the closure of some critical orbit.

37. Let S be a Riemann surface, let $f: S \to S$ be a $non-constant$ holomorphic mapping, and let $f^{\circ n}: S \to S$ be its $n-fold$ iterate. Fixing some point $z_0 \in S$, we have the following basic dichotomy: If there exists some neighborhood U of z_0 so that the sequence of iterates $\{f^{\circ n}\}$ restricted to U forms a normal family, then we say that z_0 is a **regular** or **normal** point, or that z_0 belongs to the **Fatou set** of f. Otherwise, if no such neighborhood exists, we say that z_0 belongs to the **Julia set** $J = J(f)$.

38. An ideal Q is $P-$**primary** (or is a primary ideal **belonging** to P) if Q is primary and $P = rad\ Q$.

39. Soit A un anneau linéairement topologisé.
    (i) Pour que $x \in A$ soit topologiquement nilpotent, if faut et il suffit que pour tout idéal ouvert J de A, l'image canonique de x dans A/J soit nilpotente. L'ensemble $\mathfrak{T}$ des éléments topologiquement nilpotents de A est un idéal.
    (ii) Supposons en outre que A soit préadmissible, et soit J un idéal de définition de A. Pour que $x \in A$ soit topologiquement nilpotent, il faut et il suffit que son image canonique dans A/J soit nilpotent; l'idéal $\mathfrak{T}$ est l'image réciproque du nilradical de A/J et est donc ouvert.

CHAPTER 5

1. *A presheaf $\mathfrak{F}$ on a topological space X is a **sheaf** if it satisfies the following supplementary conditions:*
   (a) if U is an open set, if $\{V_i\}$ is an open covering of U, and if $s \in \mathfrak{F}(U)$ is an element such that $s|_{V_i} = 0$ for all i, then $s = 0$;
   (b) if U is an open set, if $\{V_i\}$ is an open covering of U, and if we have elements $s_i \in \mathfrak{F}(U)$ for each i, with the property that for each $i, j, s_i|_{V_i \cap V_j} = s_j|_{V_i \cap V_j}$, then there is an element $s \in \mathfrak{F}(U)$ such that $s|_{V_i} = s_i$ for each i.

2. *Soient A un anneau admissible, J un idéal de définition de A. Alors J est contenu dans le radical de A.*

3. *A pair (S, v) consisting of a Riemann surface S and a **ramification function** $v: S \rightarrow \{1, 2, 3, \cdots\}$ which takes the value $v(w) = 1$ except at isolated points will be called a Riemann surface **orbifold**.*

4. *Let $F: X \rightarrow Y$ be a morphism of schemes, and suppose that Y can be covered by open subsets U_i, such that for each i, the induced map $f^{-1}(U_i) \rightarrow U_i$ is an isomorphism. Then F is an isomorphism.*

5. *A subgroup P of a group G is said to be a **Sylow** $\boldsymbol{p - subgroup}$ (p prime) if P is a maximal $p - $subgroup of G (that is, $P < H < G$ with H a $p - $group implies $P = H$).*

6. *If A is a subring of a field K, then the integral closure of A in K is the intersection of all valuation rings of K which contain A.*

7. An **inverse system** of abelian groups is a collection of abelian groups A_n, for each $n \geq 1$, together with homomorphisms $\varphi_{n'n}: A_{n'} \to A_n$ for each $n' \geq n$, such that for each $n'' \geq n' \geq n$ we have $\varphi_{n''n} = \varphi_{n'n} \circ \varphi_{n''n'}$.

8. Every attracting periodic orbit is contained in the Fatou set. In fact the entire basin of attraction Ω for an attracting periodic orbit is contained in the Fatou set. However the boundary $\partial\Omega$ is contained in the Julia set, and every repelling periodic orbit is contained in the Julia set.

9. Suppose M is a smooth manifold and $S \subset M$ is a submanifold (immersed or embedded). A **vector field along** S is a continuous map $N: S \to TM$ with the property that $N_P \in T_P M$ for each $P \in S$. (Note the difference between this and a vector field **on** S, which would have to satisfy $N_P \in T_P S$ at each point.) A vector $N_P \in T_P M$ at some point $P \in S$ is said to be **transverse to** S if $T_P M$ is spanned by N_P and $T_P S$. Similarly, a vector field N along S is transverse to S if N_P is transverse to S at each $P \in S$.

10. All submodules of a free R − module are free and all finitely generated torsion − free R − modules are free, provided that the ring R is a PID.

11. If A is a ring, I an ideal, and M an A − module, then $\text{depth}_I M$ is the maximum length of an M − regular sequence $x_1, \ldots, x_r$, with all $x_i \in I$.

12. Let $G(\hat{\mathbb{C}})$ be the group of all conformal automorphisms of $\hat{\mathbb{C}}$. Then every conformal automorphism g of $\mathbb{C}$ can be expressed

as a **fractional linear transformation** or **Möbius transformation** $g(z) = (az + b)/(cz + d)$, where the coefficients are complex numbers with determinant $ad - bc \neq 0$. Every non-identity automorphism of $\hat{\mathbb{C}}$ either has two distinct fixed points or one double fixed point in $\hat{\mathbb{C}}$. In general, two non-identity elements of $G(\hat{\mathbb{C}})$ commute if and only if they have precisely the same fixed points: the only exceptions to this statement are provided by pairs of commuting involutions.

13. If R is any ring, then the **center** of R is the set

$C = \{c \in R; cr = rc \text{ for all } r \in R\}.$

14. The fundamental group of the circle is infinite cyclic.

15. Let M be an R-module. A torsion-free element

$0 \neq x \in M$ is said to be **primitive** if $x = ry$ for some $y \in M$ and $r \in R$ implies that r is a unit of R.

16. Let p be the smallest prime dividing $|G|$. Then any subgroup

of G of index p is normal.

17. A scheme X_0 over a field k is **rigid** if it has no infinitesimal

deformations.

18. Let F be a field and let $\{f_i(X)\}_{i=0}^{\infty}$ be any subset of $F[X]$ such

that $\deg(f_i(X)) = i$ for each i. Then $\{f_i(X)\}_{i=0}^{\infty}$ is a basis of $F[X]$ as an F-module.

19. *A component U of the Fatou set* $\hat{\mathbb{C}} \backslash J(f)$ *is called a* **Herman**

   ring if U is conformally isomorphic to some annulus
   $\mathcal{A}_r = \{z; 1 < |z| < r\}$, *and if f* (*or some iterate of f*) *corresponds to an irrational rotation of this annulus.*

20. ***Theorem of Maschke***

   If G is a finite group and K is a field such that the characteristic of K does not divide the order of G and V is a representation of G over K, then V is completely irreducible.

21. *If G is any group, then the* **exponent** *of G is the smallest*

   natural number n such that $a^n = e$ *for all* $a \in G$. *If no such n exists, we say that G has infinite exponent.*

22. ***Theorem of Weierstrass***

   If a sequence of holomorphic functions converges uniformly, then their derivatives also converges uniformly, and the limit function is itself holomorphic.

CHAPTER 6

1. *A separable Hausdorff space M of dimension n is said to have*

   *a **differentiable structure** of class $k > 0$ if it has the following properties:*
   (i) Each point of M has an open neighborhood homeomorphic with an open subset in $\mathbb{R}^n$ the (number)space of n real variables, that is, there is a finite or countable open covering $\{U_\alpha\}$ and, for each α a homeomorphism $u_\alpha : U_\alpha \to \mathbb{R}^n$ of U_α onto an open subset in $\mathbb{R}^n$;
   (ii) For any two open sets U_α and U_β with non$-$empty intersection the map $u_\beta \circ u_\alpha^{-1} : u_\alpha(U_\alpha \cap U_\beta) \to \mathbb{R}^n$ is of class k (that is, it possesses continuous derivatives of order k) with non$-$vanishing Jacobian.

2. *If f is post critically finite, then every orbit of f is either repelling or superattracting.*

3. *A **valuation domain** is an integral domain R such that for all $a, b \in R$ either $a|b$ or $b|a$.*

4. *Every immediate attracting basin is either simply connected or infinitely connected.*

5. *Suppose $v_0, \dots, v_p$ are any $p + 1$ points in some Euclidean space $\mathbb{R}^n$. They are said to be in **general position** if they are not contained in any $(p - 1) -$ dimensional affine subspace.*

6. *Un anneau préadmissible noethérien admet un plus grand*

   idéal de définition.

7. *Let (M, g) be a Riemannian manifold. A smooth vector field*

   *Y on M is called a **Killing field** for g if g is invariant under the flow of Y.*

8. *If a compact metric space X is locally connected, then it is*

   locally path connected. Hence every connected component of X is actually path connected.

9. *Let F be an extension field of K. A **transcendence base***

   *(or **basis**) of F over K is a subset S of F which is algebraically independent over K and is maximal (with respect to set − theoretic inclusion) in the set of all algebraically independent subsets of F.*

10. *If a group G acts on a set S, then this action induces a homomorphism $G \rightarrow A(S)$, where $A(S)$ is the group of all permutations of S.*

11. *Let f be a holomorphic map from a Riemann surface to itself, and let X be a compact f − invariant subset. The map f is **hyperbolic** on X if f is expanding on X or contracting on X, or if X is the disjoint union of a compact f − invariant subset X^+ on which f is expanding and a compact f − invariant subset X^- on which f is contracting.*

12. Suppose X is a set, and we are given a transitive action of a Lie group G on X such that the isotropiy group of a point $p \in X$ is a closed Lie subgroup of G. Then X has a unique smooth manifold structure such that the given action is smooth.

13. Dans un anneau linéairement topologisé A on dit qu'un idéal J est un **idéal de définition** si J est ouvert et si, pour tout voisinage V de 0, il existe un entier $n > 0$ tel que $J^n \subset V$ (ce qu'on exprime, par abus de language, en disant que la suite (J^n) tend vers 0).

14. Let H be a subgroup of G and let $a, b \in G$. Then
    (i) $aH = bH$ if and only if $a^{-1}b \in H$, and
    (ii) $Ha = Hb$ if and only if $ab^{-1} \in H$.

15. Suppose $D \subset TM$ is a smooth distribution. An immersed submanifold $N \subset M$ is called an **integral manifold** of D if $T_P N = D_P$ at each point $P \in N$.

16. Let H and K be subgroups of a group G. Then HK is a subgroup of G if and only if $HK = KH$.

17. A pair of module homomorphisms, $A \xrightarrow{f} B \xrightarrow{g} C$, is said to be **exact** at B provided $\operatorname{Im} f = \operatorname{Ker} g$. A finite sequence of module homomorphisms, $A_0 \xrightarrow{f_1} A_1 \xrightarrow{f_2} A_2 \xrightarrow{f_3} \cdots \xrightarrow{f_{n-1}} A_{n-1} \xrightarrow{f_n} A_n$, is exact provided $\operatorname{Im} f_i = \operatorname{Ker} f_{i+1}$ for $i = 1, 2, \ldots, n - 1$. An infinite sequence of module homomorphisms, $\cdots \xrightarrow{f_{i-1}} A_{i-1} \xrightarrow{f_i} A_i \xrightarrow{f_{i+1}} A_{i+1} \xrightarrow{f_{i+2}} \cdots$ is exact provided $\operatorname{Im} f_i = \operatorname{Ker} f_{i+1}$ for all $i \in \mathbb{Z}$.

18. If z_0 belongs to the Julia set, then no sequence of iterates of f can converge uniformly throughout a neighborhood of z_0.

19. **The Adjoint Representation**

Let G be a Lie group. For any $g \in G$, the conjugate map $C_g: G \to G$ given by $C_g(h) = ghg^{-1}$ is a Lie group homomorphism. We let $Ad(g) = (C_g)_* : \mathfrak{g} \to \mathfrak{g}$ denote its induced Lie algebra homomorphism. Because $C_{g_1 g_2} = C_{g_1} \circ C_{g_2}$ for any $g_1, g_2 \in G$, it follows immediately that $Ad(g_1 g_2) = Ad(g_1) \circ Ad(g_2)$, and $Ad(g)$ is invertible with inverse $Ad(g^{-1})$. We call the map $Ad: G \to GL(\mathfrak{g})$ the **adjoint representation** of G.

20. Every compact complex manifold of dimension 1 (i.e., a compact riemann surface) is projective algebraic.

21. On dit qu'un morphisme $f: X \to Y$ de préschémas formels localement noethériens est une **immersion fermée** si elle se factorise en $X \xrightarrow{g} Z \xrightarrow{j} Y$, où g est un isomorphisme de X sur un sous − préschéma fermé Z de Y et j l'injection canonique.

22. Let $f: X \to Y$ be a dominant morphism of integral schemes of finite type over an algebraically closed field k of characteristic 0. Then there is a nonempty open set $U \subseteq X$ such that $f: U \to Y$ is smooth.

23. A ring R is **subdirectly irreducible** if the intersection of all nonzero ideals of R is nonzero.

24. If A is a ring, then evry A − module is isomorphic to a submodule of an injective A − module.

25. We say a scheme X is **regular in codimension one** (or sometimes **nonsingular in codimension one**) if every local ring O_x of X of dimension one is regular.

26. There exists one and, up to multiplication by a contant, only one Riemannian metric on the half − plane H which is

invariant under every conformal automorphism of H.

27. *Let (X, d) be a metric space . A map $G: X \to X$ is said to be a*
    contraction *if there is a constant $\lambda < 1$ such that*
    $d\big(G(x), G(y)\big) \le \lambda d(x, y)$ *for all $x, y \in X$.*

28. *Two finitely generated torsion modules over a PID are*
    isomorphic if and only if they have the same chain of invariant
    ideals.

29. *A **smooth function element** on a smooth manifold M is*
    an ordered pair (f, U), where U is an open subset of M and
    $f: U \to \mathbb{R}$ is a smooth function.

30. ***Nakai − Moishezon Criterion***
    A divisor D on the surface X is ample if and only if $D^2 > 0$ and
    $D \cdot C > 0$ for all irreducible curves C in X.

31. *A region R is called **simply − connected** if any simple*
    closed curve which lies in R can be shrunk to a point without
    leaving R. A region R which is not simply − connected is called
    multiply − connected.

32. *If $\alpha: M^m \to N^n$ is a smooth map between manifolds of*
    dimensions $m \ge n$, and if $y \in N$ is a regular value, then the set
    $\alpha^{-1}(y) \subset M$ is a smooth manifold of dimension $m - n$.

33. *Let W be the germ at w of a normal two − dimensional*
    complex analytic space with singularities at w, and let
    *$\psi: M \to W$ be a resolution of W. Then we define the **exceptional***
    ***set** E of the resolution to be the set $\psi^{-1}(w)$.*

34. *In a compact and orientable Riemannian manifold an*

infinitesimal affine collineation is a motion.

35. If A is a module over a commutative ring R and $a \in A$,
then $\mathcal{O}_a = \{r \in R; ra = 0\}$ is an ideal of R. If $\mathcal{O}_a \neq 0, a$ is said
to be a **torsion element** of A. If R is an integral domain, then
the set T(A) of all torsion elements of A is a submodule of A,
called the **torsion submodule**.

36. Every integrable distribution is involutive.

37. Un **espace annelé** (resp. **topologiquement annelé**)
est un couple $(X, \mathcal{A})$ formé d'un espace topologique X et d'un
faisceau d'anneaux (non nécessairement commutatifs)
(resp. d'un faicseau d'anneaux topologiques) $\mathcal{A}$ sur X; on dit que
X est l'espace topologique **sous $-$ jacent** à l'espace annelé
$(X, \mathcal{A})$, et $\mathcal{A}$ le **faisceau structural**. Ce dernier se note $\mathcal{O}_X$, et sa
fibre en un point $x \in X$ se note note $\mathcal{O}_{X,x}$ ou simplement $\mathcal{O}_x$
lorsqu'il n'en résulte pas de confusion.

38. **Schur's Lemma**
(i) Let M be a simple $R -$ module. Then the ring $End_R(M)$ is a
division ring.
(ii) If M and N are simple $R -$ modules, then $Hom_R(M, N) \neq \langle 0 \rangle$
if and only if M and N are isomorphic.

39. (i) A ring R is an **integral domain** if it is a commutative
ring with identity such that R has no zero divisors.
(ii) A ring with identity is a **division ring** if every nonzero
element of R has a multiplicative inverse.
(iii) A **field** is a commutative division ring.

40. Let G be a linear algebraic $k -$ group such that the Lie
algebra of every reductive subquotient of G has only a finite
number of conjugacy classes of nilpotent elements. Then the

maximal connected unipotent k − subgroups of G are the unipotent radicals of the minimal parabolic k − subgroups of G.

41. A complex number is **algebraic** if it is a root of a polynomial with integer coefficients, and a subfield $F \subseteq \mathbb{C}$ is said to be algebraic if every element of F is algebraic. If F is algebraic, the **integers** of F are those elements of F that are roots of a monic polynomial with integer coefficients.

42. If $D \subset \mathbb{R}^n$ is a domain of integration, then every bounded continuous function on D is integrable over D.

43. Let K be a field and $f \in K[x]$. The equation $f(x) = 0$ is **solvable by radicals** if there exists a radical extension F of K and a splitting field E of f over K such that $K \subset E \subset F$.

44. Let $\mathfrak{g}$ be a finite − dimensional Lie algebra. The connected Lie groups whose Lie algebras are isomorphic to $\mathfrak{g}$ are (up to isomorphism) precisely those of the form G/Γ, where G is the simply connected Lie group with Lie algebra $\mathfrak{g}$, and Γ is a discrete central subgroup of G.

45. A space X is **arcwise connected** if there is a topological embedding of $[0, 1]$ into X which joins any two given distinct points.

46. Suppose that M is a compact manifold without boundary and that $f \colon M \to \mathbb{R}$ is a smooth function. If the interval $[a_1, a_2]$ contains no critical value of f then the manifolds with boundary $M_{(-\infty, a_1]}$ and $M_{(-\infty, a_2]}$ are diffeomorphic.

CHAPTER 7

1. *If M is a smooth n − manifold, a smooth, compact, embedded*
   n − dimensional submanifold with boundary $D \subset M$ is called a
   regular domain *in M.*

2. ***A theorem of Kostant***
   Let K be algebraically closed of characteristic 0.
   (a) If $\mathfrak{g}$ is a semi − simple Lie algebra over K, then $\mathfrak{g}$ has only a
   finite number of conjugacy classes of nilpotent elements.
   (b) Let G be a semi − simple linear algebraic group over K. Then
   G has only a finite number of conjugacy classes of unipotent
   elements.

3. *We say that the R − module M is **finitely generated** if*
   *$M = \langle S \rangle$ for some finite subset S of M. M is **cyclic** if $M = \langle m \rangle$ for*
   some element $m \in M$. If M is finitely generated then let $\mu(M)$
   denote the minimal number of generators of M. If M is not
   finitely generated, then we will define $\mu(M) = \infty$. We will call
   *$\mu(M)$ the **rank** of M.*

4. *For any manifold M, the projection $\pi: M \times \mathbb{R} \to M$ and the*
   zero − section $M \to M \times \mathbb{R}$ induce mutually inverse isomorphisms
   on cohomology.

5. *Let F be an extension field of K. A transcendence base S of*
   *F over K is called a **separating transcendence base** of F over*
   K if F is separable algebraic over K(S). If F has a separating
   *transcendence base over K, then F is said to be **separately***
   ***generated** over K.*

6. *The following conditions on an $R-$ module P are equivalent.*
   (1) Every short exact sequence of $R-$ modules
   $0 \to M_1 \to M \to P \to 0$ *splits.*
   (2) There is an $R-$ module P' such that $P \oplus P'$ is a free
   $R-$ *module.*
   (3) For any $R-$ module N and any surjective $R-$ module
   homomorphism $\psi: M \to P$, the homomorphism
   $\psi_*: Hom_R(N, M) \to Hom_R(N, P)$ is surjective. That is, for any
   homomorphism $\alpha: N \to P$, there exists a homomorphism $\beta: N \to M$,
   such that $\alpha = \psi \circ \beta$.
   (4) For any surjective $R-$ module homomorphism $\phi: M \to N$, the
   homomorphism $\phi_*: Hom_R(P, M) \to Hom_R(P, N)$ is surjective.

7. *A symplectic manifold (M, ω) together with a smooth*
   *function $H \in C^\infty(M)$ is called a **Hamiltonian system**. The*
   *function H is called the **Hamiltonian** of the system; the flow of*
   *the Hamiltonian vector filed X_H is called its **Hamiltonian flow**,*
   *and the integral curves of X_H are called the **trajectories** or*
   ***orbits** of the system.*

8. ***Zariski's Main Theorem***
   Let $T: X \to Y$ be a birational transformation of projective
   varieties, and assume that X is normal. If P is a fundamental
   point of T, then the total transform $T(P)$ is connected and of
   dimension greater or equal to 1.

9. *Let M be an $R-$ module and $S \subseteq M$ a submodule. Then S is*
   *said to be **complemented** if there exists a submodule $T \subseteq M$*
   such that $M \cong S \oplus T$.

10. *If R is a Noetherian local ring with maximal ideal M, then*
    $$\bigcap_{n=1}^{\infty} M^n = 0.$$

11. *A smooth manifold endowed with a transitive smooth action by a Lie group G is called a **homogeneous G − space** (or a **homogeneous space** or **homogeneous manifold** if it is not important to specify the group).*

12. *Pour qu'un sous − espace Y d'un espace topologique X soit irréductible, il faut et il suffit que son adhérence $\bar{Y}$ soit irréductible. En particulier, tout sous − espace qui est l'adhérence $\overline{\{x\}}$ d'un sous − espace réduit à un point est irréductible; nous exprimerons la relation $y \in \overline{\{x\}}$ (équivalente à $\overline{\{y\}} \subset \overline{\{x\}}$) en disant que y est **spécialisation de** x ou que x est une **générisation de** y. Lorsqu'il existe dans un espace irréductible X un point x tel que $X = \overline{\{x\}}$, nous dirons que x est **point générique** de X. Tout ouvert non vide de X contient alors x et et tout sous − espace contenant x admet x pour point générique.*

13. *Let $\widetilde{M}$ and M be topological spaces, and let $\pi: \widetilde{M} \to M$ be a (topological)covering map. A **covering transformation** (or **deck transformation**) of π is a homeomorphism $\varphi: \widetilde{M} \to \widetilde{M}$ such that $\pi \circ \varphi = \pi$.*

14. *Let V be a rational $GLn(k) −$ module and V_1 a submodule of V. Let $V_2 = V/V_1$ be the factor space. Then the weights of V are the weights of V_1 together with the weights of V_2.*

15. *A **Prüfer domain** is an integral domain in which every finitely generated ideal is invertible.*

16. *Suppose ρ is a rational representation of $GLn(k)$ and $g \in GLn(k)$ is unipotent. Suppose Char $k = p$. Then $\rho(g)$ is unipotent.*

17. *Let I be an ideal in a commutative ring R. The* **radical**
    (or **nilradical***)of I, denoted Rad I, is the ideal $\cap P$, where the
    intersection is taken over all prime ideals P which contain I.
    If the set of prime ideals containing I is empty, then Rad I is
    defined to be R.*

18. *Let $f: X \to Y$ be a morphism of varieties over k. Assume
    that Y is regular, X is Cohen $-$ Macaulay, and that every fibre
    of f has dimension equal to $\dim X - \dim Y$. Then f is flat.*

19. *A $GLn(k) -$ module is a finite dimensional vector space
    V over k, with an action $GLn(k) \times V \to V, (g, v) \mapsto gv$, satisfying
    the actions:
    (i) $v \mapsto gv$ is linear
    (ii) $(g_1 g_2)v = g_1(g_2 v)$
    (iii) $1v = v$*

20. **Wedderbum's First Theorem**
    *Let A be a finite dimensional algebra over k (algebraically
    closed). If A is semisimple, then A can be expressed uniquely as
    $A = A_1 \oplus \cdots \oplus A_n$ of simple algebras A_i.*

21. *Let M be a smooth $n -$ manifold, and let (U, φ) be a smooth
    chart on M. If S is a subset of U such that $\varphi(S)$ is a $k -$ slice of
    $\varphi(U)$, then we say simply that S is a $k -$ lice of U.*

22. **Wedderbum's Second Theorem**
    *Each simple algebra over k is isomorphic to a complete matrix
    algebra over k.*

23. *An* **affine scheme** *is a locally ringed space $(X, \mathcal{O}_X)$ which
    is isomorphic (as a locally ringed space) to the spectrum of
    some ring. A* **scheme** *is a locally ringed space $(X, \mathcal{O}_X)$ in which
    every point has an open neighborhood U such that the*

topological space U, together with the restricted sheaf $\mathcal{O}_X|_U$, is an **affine scheme**. We call X the **underlying topological space** of the scheme $(X, \mathcal{O}_X)$, and $\mathcal{O}_X$ its **structure sheaf**.

24. If P is a projective R − module, then $\eta: P \to P^{**}$ is injective. If P is also finitely generated, then P is reflexive.

25. A smooth group homomorphism $\alpha: (\mathbb{R}, +) \to G$ is called a **one parameter subgroup** of G.

26. Let R be a commutative ring, let M be a finitely generated R − module, and let $S \subseteq M$ be a subset. If $|S| > \mu(M) = rank(M)$, then S is not R − linearly independent.

27. Let V, W be Banach spaces. A linear map $T: V \to W$ is said to be **compact** if the image $T(B)$ of the unit ball $B = B(0; 1) \subset V$ has compact closure in W.

28. **The Fundamental Theorem of Algebra**
    The field of complex numbers is algebraically closed.

29. Let K be a field. the **Galois group** of a polynomial $f \in K[x]$ is the group $Aut_K F$, where F is a splitting field of f over K.

30. **Contraction Lemma**
    Let X be a complete metric space. Then every contraction $G: X \to X$ has a unique fixed point, i.e., a point $x \in X$ such that $G(x) = x$.

31. An **abstract variety** is an integral separated scheme of finite type over an algebraically closed field k. If it is proper over k, we will also say it is **complete**.

32. Suppose U and V are open subsets of $\mathbb{R}^n$, and $F: U \to V$ is a smooth map. If $DF(p)$ is nonsingular at some point $p \in U$, then

there exist connected neighborhoods $U_0 \subset U$ of p and $V_0 \subset V$ of $F(p)$ such that $F|_{U_0} : U_0 \to V_0$ is a diffeomorphism.

33. *A function $f : V \to W$ between vector spaces over a field k of positive characteristic p is **semilinear** if f is an abelian group homomorphism and $f(\alpha v) = \alpha^p v$.*

34. *An $R - $ module P is a finitely generated projective $R - $ module if and only if P is a direct summand of a finitely generated free $R - $ module.*

35. *Let $f : E \to F$ be a linear transformation of (left) vector spaces over a division ring D. The **rank** of f is the dimension of $\text{Im } f$ and the **nullity** of f is the dimension of $\text{Ker } f$.*

36. ***Gluing Lemma***
Let X and Y be topological spaces, and let $K_1, \ldots, K_n$ be finitely many closed subsets of X whose union is X. Suppose that we are given continuous maps $f_i : K_i \to Y, i = 1, \ldots, n$, that agree on overlaps: $f_i|_{K_i \cap K_j} = f_j|_{K_i \cap K_j}$. Then there exists a unique continuous map $f : X \to Y$ whose restriction to each K_i is equal to f_i.

37. *On dit qu'un espace topologique X est **noethérien** si l'ensemble des ouverts de X vérifie la condition **maximale**, ou, ce qui revient au même, si l'ensemble des fermés de X vérifie la condition **minimale**. On dit que X est **localement noethérien** si tout $x \in X$ admet un voisinage qui est un sous $-$ espace noethérien.*

CHAPTER 8

1. *Soient A, B deux anneaux locaux, $\mathfrak{m}, \mathfrak{n}$ leurs idéaux maximaux, et supposons B noethérien. Soient $\varphi: A \to B$ un homomorphism local, M un B — module de type fini. Si M est un A — module quasi — fini, les topologies $\mathfrak{m}$ — préadiques et $\mathfrak{n}$ — préadiques sur M sont identiques, donc séparées.*

2. *A **Dedekind domain** is an integral domain R in which every ideal ($\neq R$) is the product of a finite number of prime ideals.*

3. *Let H be a closed subgroup of the Lie group G. Then the right action of H on G is proper and free.*

4. *Suppose M and N are oriented positive — dimensional manifolds, and $F: M \to N$ is a local diffeomorphism. We say that F is **orientation — preserving** if for each $p \in M, F_*$ takes oriented bases of $T_P M$ to oriented bases of $T_{F(P)} N$, and **orientation — reversing** if it takes oriented bases of $T_P M$ to negatively oriented bases of $T_{F(P)} N$.*

5. *Let M be a connected smooth manifold, and suppose that the fundamental group of M has no subgroup of index 2. Then M is orientable. In particular, this is the case if M is simply connected.*

6. *Let $B = k[x]$ with k a field; a k — **automorphism** of B is a ring homomorphism $\varphi: B \to B$ that is the identity on k and is an automorphism of B (that is, one — to — one and onto).*

7. *Tout S — préschéma qui est affine au — dessus de S est séparé au — dessus de S (autrement dit, est un S — schéma).*

8. Let $\{f_n\}$ be a family of holomorphic (or meromorphic) functions defined on some open set U. Then $\{f_n\}$ converges **locally uniformly** to a function f if the sequence converges uniformly on every compact subset of U.

9. **Hurwitz's Theorem**
   If $f_n \to f$ locally uniformly and $f_n(z) \neq 0$ for all z, then either $f \neq 0$ for all z, or f is constant.

10. A point $z \in X$ with $f^n(z) = z$ is called **periodic**. The smallest such n is the **period** of z.) Such a periodic point is called
   - **attracting** if $0 < |(f^n)'(z)| < 1$;
   - **superattracting** if $|(f^n)'(z)| = 0$;
   - **repelling** if $|(f^n)'(z)| > 1$;
   - **indifferent** (or **neutral**)if $|(f^n)'(z)| = 1$.

11. Soit S un schéma affine; pour qu'un S − préschéma X soit affine au − dessus de S, il faut et il suffit que X soit un schéma affine.

12. Nous dirons qu'un morphisme $f : X \to Y$ de préschémas est **affine** s'il définit X comme un préschéma affine au − dessus de Y.

13. A space X is simply − connected if and only if there is a unique homotopy class of paths connecting any two points in X.

14. A **covering space** of a space X is a space $\tilde{X}$ together with a map $\rho : \tilde{X} \to X$ satisfying the following condition: there exists an open cover $\{U_\alpha\}$ of X such that for each α, $\rho^{-1}(U_\alpha)$ is a disjoint union of open sets in $\tilde{X}$, each of which is mapped

homeomorphically onto U_α by ρ. The empty disjoint union is allowed, so ρ need not be surjective.

15. Soit $f: X \to Y$ un morphisme étale. Si Y est normal, X l'est, la réciproque étant vraie si f est surjective.

16. A space X is **paracompact** if it is Hausdorff and if every covering $\mathcal{A}$ of X has a locally finite open refinement $\mathcal{B}$ that covers X.

17. Every Lie group has precisely two left $-$ invariant orientations, corresponding to the two orientations of its Lie algebra.

18. An indexed family of sets $\{A_\alpha\}$ is said to be **locally finite** if each point x of X has a neighborhood that intersects A_α for only finitely many values of α.

19. If F is a field and G is a finite subgroup of the multiplicative group of nonzero elements of F, then G is a cyclic group. In particular, the multiplicative group of all nonzero elements of a finite field is cyclic.

20. Let $(x_\alpha)_{\alpha \in J}$ be a net in X. We say that x is an **accumulation point** of the net $(x_\alpha)_{\alpha \in J}$ if for each neighborhood U of x, the set of those α for which $x_\alpha \in U$ is cofinal in J.

21. Let G be a topological group; let A and B be subsets of G. If A is closed in G and B is compact, then $A \cdot B$ is closed in G.

22. A collection $\mathcal{A}$ of subsets of X satisfies the **countable intersection condition** (c.i.c) if every countable intersection of elements of $\mathcal{A}$ is nonempty.

23.**Regular Level Set Theorem**

    Every regular level set of a smooth map is a closed embedded submanifold whose codimension is equal to the dimension of the range.

24.*A space X is said to be **contractible** if the identity map $i_X: X \to X$ is homotopic to a constant map.*

25.*If M is a connected smooth manifold, any two points of M can be joined by a piecewise smooth curve segment.*

26.*Let F be a field and $f \in F[x]$ a polynomial of positive degree. f is said to **split** over F (or to **split in** $F[x]$) if f can be written as a product of linear factors in $F[x]$; that is, $f = u_0(x - u_1)(x - u_2) \cdots (x - u_n)$ with $u_i \in F$.*

27.*Let X be a (nonempty) compact Hausdorff space. If every point of X is a limit point of X, then X is uncountable.*

28.*Let K be a field and $f \in K[x]$ a polynomial of positive degree. An extension field F of K is said to be a **splitting field over** K **of the polynomial** f if f splits in $F[x]$ and $F = K(u_1, \ldots, u_n)$ where $u_1, \ldots, u_n$ are the roots of f in F.*

29.**Existence of Bump Functions**

    Let M be a smooth manifold. For any closed set $A \subset M$ and any open set U containing A, there exists a smooth bump function for A supported in U.

30.*Let $p: E \to B$ be a map. If f is a continuous mapping of some space X into B, a **lifting** of f is a map $\tilde{f}: X \to E$ such that $p \circ \tilde{f} = f$.*

31. **Global Frobenius Theorem**
    Let D be an involutive distribution on a smooth manifold M. The collection of all maximal connected integral manifolds of D forms a foliation of M.

32. *Let K be a field and S a set of polynomials of positive degree in K[x]. An extension field F of K is said to be a* **splitting field over** *K* **of the set** *S* **of polynomials** *if every polynomial in S splits in F[x] and F is generated over K by the roots of all the polynomials in S.*

33. *Let $\mathcal{F}$ be a foliation on a smooth manifold M. The collection of tangent spaces to the leaves of $\mathcal{F}$ forms an involutive distribution on M.*

34. *A map $h: X \rightarrow Y$ is said to be* **inessential** *if h is homotopic to a constant map. Otherwise, it is said to be* **essential**.

35. *Let Y be a locally complete intersection subscheme of a nonsingular variety X over k. Then:*
    (a) Y is Cohen − Macaulay;
    (b) Y is normal if and only if it is regular in codimension 1.

36. *On dit qu'un espace topologique X est* **équidimensionnel** *si toutes ses composantes irréductibles ont même dimension.*

37. *Let X be compact Hausdorff; let $x \in X$. The intersection of all those sets A containing x which are both open and closed in X equals the components of X containing x.*

CHAPTER 9

1. *A scheme X with a morphism to another scheme S is a*
    group scheme over *S if there is a section $e : S \to X$ (the identity)*
    and a morphism $\rho : X \to X$ over S (the inverse) and a morphism
    $\mu : X \times X \to X$ over S (the group operation) such that
    (1) the composite $\mu \circ (id \times \rho) : X \to X$ is equal to the projection
    $X \to S$ followed by e, and
    (2) the two morphisms $\mu \circ (\mu \times id)$ and $\mu \circ (id \times \mu)$ from
    $X \times X \times X \to X$ are the same.

2. ***The Riemannian Density***
    Let (M, g) be a Riemannian manifold with or without boundary.
    There is a unique smooth positive density μ on M, called the
    Riemannian density, *with the property that $\mu(E_1, \ldots, E_n) = 1$*
    for any local orthonormal frame (E_i).

3. *A sequence $(x_1, x_2, \ldots)$ of points of X is said to **converge***
    to the point x of X if for every neighborhood U of x there exists
    a positive integer N such that x_i lies in U for all $i \geq N$.

4. ***Invariance of Corner Points***
    Let M be a smooth $n-$ manifold with corners, and let $p \in M$. If
    $\varphi(p)$ is a corner point for some smooth chart with corners (U, φ),
    then the same is true for every such chart whose domain contains
    p.

5. *Let G be a finite group. A field F is called **excellent for** G*
    *or simply **excellent** if it is algebraically closed of characteristic*
    zero or prime to the order of G.

6. *The arithmetic genus of a nonsingular projective surface*

is a birational invariant.

7. *An algebraic extension field F of K is **normal** over K (or a* *normal extension) if every irreducible polynomial in K[x] that* *has a root in F actually splits in F[x].*

8. **Functions with Vanishing Differentials**
   If f is a smooth real − valued function on a smooth manifold M, *then df = 0 if and only if f is constant on each component of M.*

9. *Let R and S be rings. An abelian group M is an* *(R, S) − **bimodule** if M is both a left R − module and a right* *S − module, and the compatibility condition r(ms) = (rm)s is* *satisfied for every r ∈ R, m ∈ M, and s ∈ S.*

10. *Let G be a finite group and F a good field for G. Then G is* *abelian if and only if G has n distinct irreducible* *F − representations.*

11. *Let M be a topological space. A (**real**) **vector bundle of*** ***rank k over** M is a topological space E together with a* *surjective continuous map π: E → M satisfying:*
    *(i) for each p ∈ M, the set $E_p = \pi^{-1}(p) \subset E$ (called the **fiber** of* *E over p) is endowed with the structure of a k − dimensional* *real vector space.*
    (ii) for each p ∈ M, there exist a neighborhood U of p in M and *a homomorphism $\phi: \pi^{-1}(U) \to U \times \mathbb{R}^k$ (called a **local*** ***trivialization** of E over U), such that $\pi|_{\pi^{-1}(U)} = \pi_1 \circ \phi$, where* *$\pi_1: U \times \mathbb{R}^k \to U$ is the projection on the first factor; and such* *that for each q ∈ U, the restriction of φ to E_p is a linear* *isomorphism from E_p to $\{q\} \times \mathbb{R}^k \cong \mathbb{R}^k$.*

12. *If S is a complemented submodule of a finitely generated*

R − module (R a PID), then any basis for S extends to a basis for M.

13. An algebra A with identity over a field K is said to be **central simple** if A is a simple K − algebra and the center of A is precisely K.

14. Suppose X is a set, and we are given a transitive action of a Lie group G on X such that the isotropy group of a point $p \in X$ is a closed Lie subgroup of G. Then X has a unique smooth manifold structure such that the given action is smooth.

15. Let R be an integral domain and let K be the quotient field of R. An ideal $I \subseteq R$ is said to be **invertible** if there are elements $a_1, \dots, a_n \in I$ and $b_1, \dots, b_n \in K$ such that
(1) $b_i I \subseteq R$ for $1 \leq i \leq n$, and
(2) $a_1 b_1 + \cdots + a_n b_n = 1$.

16. **Escape Lemma**
Let V be a smooth vector field on a smooth manifold M. If γ is an integral curve of V whose maximal domain is not all of $\mathbb{R}$, then the image of γ cannot lie in any compact subset of M.

17. An R − module M is said to be **semisimple** if it is a direct sum of simple R − modules.

18. Soit $f: X \to Y$ un S − morphisme tel que Y soit l'image fermée de X par f. Soit Z un S − schéma; si deux S − morphismes g_1, g_2 de Y dans Z sont tels que $g_1 \circ f = g_2 \circ f$, alors $g_1 = g_2$.

19. If X is a topological space and $S \subset X$ is any subset, we define the **subspace topology** (sometimes called the **relative topology**) on S by declaring a subset $U \subset S$ to be open in S if and only if there exists an open set $V \subset X$ such that $U = V \cap S$.

20. Soient A un anneau local, $\mathfrak{m}$ son idéal maximal, B une A − algèbre telle que $\mathfrak{m}B \neq B$ (ce qui a lieu par example lorsque B est un anneau local et $A \rightarrow B$ un homomorphisme local). Si B est un A − module plat, B est un A − module fidèlement plat.

21. Let M be a smooth manifold. A **Riemannian metric** on M is a smooth symmetric 2 − tensor field that is positive definite at each point.

22. **Bing Metrization Theorem**
   A space X is metrizable if and only if it is regular and has a basis that is countably locally discrete.

23. On appelle **schéma local** un schéma affine dont l'anneau A est local; il existe alors dans $X = Spec(A)$ un seul **point fermée** a, et pour tout autre point $b \in X$, on a $a \in \overline{\{b\}}$.

24. Let X be a scheme. The kernel, cokernel, and image of any morphism of quasi − coherent sheaves are quasi − coherent. Any extension of quasi − coherent sheaves is quasi − coherent. If X is noetherian, the same is true for coherent sheaves.

25. If M is a manifold, a **flow domain** for M is an open subset $D \subset \mathbb{R} \times M$ with the property that for each $p \in M$, the set $D^{(p)} = \{t \in \mathbb{R}; (t, p) \in D\}$ is an open interval containing 0.

26. Let R be a semisimple ring. Then every simple R − module is isomorphic to a submodule of R.

27. If A, B are local rings contained in a field K, we say that B **dominates** A if $A \subseteq B$ and $\mathfrak{m}_B \cap A = \mathfrak{m}_A$.

CHAPTER 10

1. ***Schroeder – Bernstein Theorem***
   If there are injections $f: A \rightarrow C$ and $g: C \rightarrow A$, then A and C have the same cardinality.

2. *Let R be an integral domain with quotient field K. A **fractional ideal** of R is a nonzero R – submodule I of K such that $aI \subset R$ for some nonzero $a \in R$.*

3. *Suppose $H \subset G$ is a Lie subgroup. The one – parameter subgroups of H are precisely those one – parameter subgroups of G whose initial tangent vectors lie in $T_e H$.*

4. *If X is a topological space and $p \in X$, a **neighborhood basis** at p is a collection $\mathcal{B}_p$ of neighborhoods of p such that every neighborhood of p contains at least one $B \in \mathcal{B}_p$.*

5. *(a) Soient f un morphisme $Y \rightarrow X$, (V_λ) un recouvrement de $f(Y)$ par des ouverts de X. Pour que f soit une immersion (resp. immersion ouverte), il faut et il suffit que sa restriction à chacun des préschémas induits $f^{-1}(V_\lambda)$ soit une immersion (resp. une immersion ouverte) dans V_λ.*
   (b) Soient f un morphisme $Y \rightarrow X$, (V_λ) un recouvrement ouvert de X. Pour que f soit une immersion fermée, il faut et il suffit que sa restriction à chacun des préschémas induits $f^{-1}(V_\lambda)$ soit une immersion fermée dans V_λ.

6. *If X and Y are topological spaces, a continuous injective map $f: X \rightarrow Y$ is called a **topological embedding** if it is a homeomorphism onto its image $f(X) \subset Y$ in the subspace topology.*

7. Let R be an integral domain. If P is a projective
   $R-module$, then P is torsion $-$ free.

8. Let S be a Riemann surface, let $f: S \to S$ be a
   non $-$ constant holomorphic mapping, and let $f^{\circ n}: S \to S$ be its
   $n-$ fold iterate. By the **grand orbit** of a point z under f we
   mean the set $GO(z, f)$ consisting of all points $z' \in S$ whose orbits
   eventually intersect the orbit of z. Thus z and z' have the same
   grand orbit if and only if $f^{\circ m}(z) = f^{\circ n}(z')$ for some choice of
   $m \geq 0$ and $n \geq 0$.

9. If X is a simply connected space, then any covering map
   $\pi: \tilde{X} \to X$ is a homeomorphism.

10. Let $f: M \to M$ be a $C_1 -$ smooth map from a smooth
    Riemannian manifold to itself, and let $X \subset M$ be a compact
    $f -$ invariant compact subset, $f(X) \subset X$. Let $Df_x: T_x M \to T_{f(x)} M$, or
    briefly Df, be the **first derivative map** at x, that is, the induced
    linear map on the tangent space at x, and let $||v||$ be the
    Riemannian norm of a vector $v \in T_x M$. By definition, the map f
    is **expanding** on X if the length of any tangent vector at a point
    of X expands exponentially under iteration of Df, that is, if there
    exists constants $c > 0$ and $k > 1$ so that $||Df^{\circ n}(v)|| \geq ck^n ||v||$ for
    every $x \in X$ and $v \in T_x M$, and every $n \geq 0$. Since X is compact, a
    completely equivalent requirement is that there exists some fixed
    $n \geq 1$ so that $||Df^{\circ n}(v)|| > ||v||$ for all non $-$ zero tangent vectors
    v over X. Similarly, f is **contracting** on X if there are constants
    $c >$ and $k < 1$ so that $||Df^{\circ n}(v)|| \leq ck^n ||v||$.

11. Si A est un anneau admissible et J est contenu dans un idéal
    de définition de A, A est séparé et complet pour la topology
    $J -$ préadique.

12. Suppose $\tilde{X}$ and X are topological spaces. A map $\pi: \tilde{X} \rightarrow X$ is called a **covering map** if $\tilde{X}$ is path connected and locally path connected, π is surjective and continuous, and each point $p \in X$ has a neighborhood U that is **evenly covered** by π, meaning that U is connected and each component of $\pi^{-1}(U)$ is mapped homeomorphically onto U by π. In this case, X is called the **base** of the covering, and $\tilde{X}$ is called a **covering space** of X.

13. Let $\phi: S \rightarrow \mathbb{P}_1$ be a meromorphic function on a Riemann surface S which takes any value $a \in \mathbb{P}_1 = \mathbb{C} \cup \{\infty\}$ exactly d times, counting multiplicities. Then S is compact.

14. A continuous map $\pi: Y \rightarrow X$ between topological spaces X and Y is called a **covering projection** if each $x \in X$ has an open neighborhood U in X such that $\pi^{-1}(U)$ is a disjoint union of open subsets of Y each of which is mapped homeomorphically onto U.

15. Let $T: X \rightarrow X'$ be a birational transformation of surfaces. Then it is possible to factor T into a finite sequence of momoidal transformations and their inverses.

16. A morphism $f: X \rightarrow Y$, with Y irreducible, is **generically finite** if $f^{-1}(\eta)$ is a finite set, where η is the generic point of Y. A morphism $f: X \rightarrow Y$ is **dominant** if $f(X)$ is dense in Y.

17. Primitive Element Theorem
Let F be a finite dimensional extension field of K.
(i) If F is separable over K, then F is a simple extension of K.
(ii) (**Artin**) More generally, F is a simple extension of K if and only if there are only finitely many intermediate fields.

18. *A **topological group** G is a group that is also a topological space, satisfying the requirements that the map of $G \times G$ into G sending $x \times y$ into $x \cdot y$, and the map of G into G sending x into x^{-1}, are continuous.*

19. *Regularity Axiom*
Given a closed set C and a point x not in C, there exist disjoint open sets containing C and x, respectively.

20. *A topological space X is a **Zariski space** if it is noetherian and every (nonempty) closed irreducible subset has a unique generic point.*

21. *If Y is a closed subscheme of an affine scheme $X = Spec\ A$, then Y is also affine, and in fact Y is the closed subscheme determined by a suitable ideal $\mathfrak{a} \subseteq A$ as the image of the closed immersion $Spec\ A/\mathfrak{a} \rightarrow Spec\ A$.*

22. *Let X be a Zariski topological space. A **constructible subset** of X is a subset which belongs to the smallest family $\mathcal{F}$ of subsets such that*
(1) every open subset is in $\mathcal{F}$,
(2) a finite intersection of elements of $\mathcal{F}$ is in $\mathcal{F}$, and
(3) the complement of an element of $\mathcal{F}$ is in $\mathcal{F}$.

23. *If M is a smooth manifold and $N \subset M$ is an immersed submanifold of positive codimension, then N has measure zero in M.*

24. *A scheme $(X, \mathcal{O}_X)$ is **reduced** if for every open set $U \subseteq X$, the ring $\mathcal{O}_X(U)$ has no nilpotent element.*

25. Let A be a ring, let $X = Spec\ A$, let $f \in A$ and let $D(f) \subseteq X$ be the open complement of $V((f))$. Then the locally ringed space $(D(f), \mathcal{O}_X|_{D(f)})$ is isomorphic to $Spec\ A_f$.

26. Let A be an abelian group. We define the **constant presheaf associated** to A on the topological space X to be the presheaf $U \longmapsto A$ for all $U \neq \emptyset$, with restriction maps the identity.

27. An algebraic set Y of pure dimension r (i.e., every irreducible component of Y has dimension r) has degree 1 if and only if Y is a linear variety.

28. For a point P on a variety X, let $\mathfrak{m}$ be the maximal ideal of the local ring $\mathcal{O}_P$. We define the **Zariski tangent space** T_PX of X at P to be the dual $k-$ vector space of $\mathfrak{m}/\mathfrak{m}^2$.

29. If f and g are regular functions on open subsets U and V of a variety X, and if $f = g$ on $U \cap V$, then the function which is f on U and g on V is a regular function on $U \cup V$. If f is a rational function on X, then there is a largest open subset U of X on which f is represented by a regular function. We say that f is **defined** at the points of U.

30. A variety Y is **normal at a point** $P \in Y$ if $\mathcal{O}_P$ is an integrally closed ring. Y is **normal** if it is normal at every point.

31. If X is a quasi $-$ affine or quasi $-$ projective variety and Y is an irreducible locally closed subset, then Y is also a quasi $-$ affine (respectively, quasi $-$ projective) variety, by virtue of being a locally closed subset of the same affine or projective space. We call this the **induced structure** on Y, and we call Y a **subvariety** of X.

32. If A is a ring, $\mathfrak{a}$ an ideal, and M an $A - module$, then
$depth_{\mathfrak{a}}M$ is the maximum length of an $M - regular$ sequence
$x_1, ..., x_r$, with all $x_i \in \mathfrak{a}$.

33. Let X be a noetherian scheme, and let P be a closed point
of X. Then the following conditions are equivalent:
(i) depth $\mathcal{O}_P \geq 2$;
(ii) if U is any open neighborhood of P, then every section of
$\mathcal{O}_X$ over $U\{P\}$ extends uniquely to a section of $\mathcal{O}_X$ over U.

34. Let $Y \subseteq X$ be a closed subscheme, where X is a scheme of
finite type over a field k. Let $D = k[t]/t^2$ be the ring of dual
numbers, and define an **infinitesimal deformation of** Y **as
a closed subscheme of** X, to be a closed subscheme
$Y' \subseteq X \times_k D$, which is flat over D, and whose closed fibre is Y.

35. If $f: X \rightarrow Y$ is a finite surjective morphism of nonsingular
varieties over an algebraically closed field k, then f is flat.

36. Let Y be defined by the equation $f(x, y) = 0$ in $\mathbb{A}^2$, and let
$P = (0, a)$ be a point of multiplicity r on Y, so that when f is
expanded as a polynomial in x and y, we have $f = f_r + higher$
terms. We say that P is an **ordinary** $r - $**fold** point if f_r is a
product of r distinct linear factors.

37._Chow's Lemma_
Let X be proper over a noetherian scheme S. Then there is a
scheme X' and a morphism $g: X' \rightarrow X$ such that X' is projective
over S, and there is an open dense subset $U \subseteq X$ such that g
induces an isomorphism of $g^{-1}(U)$ to U.

38. Let S be a graded ring, generated by S_1 as an $S_0 - algebra$,
let M be a graded $S - module$. We say that M is **quasi** $-$ **finitely**

generated if it is equivalent to a finitely generated module.

39. Let X be a reduced noetherian scheme. Then X is affine if and only if each irreducible components of X is affine.

40. A curve X is called **hyperelliptic** if its genus $g \geq 2$ and there exists a finite morphism $f: X \to \mathbb{P}^1$ of degree 2.

41. Let M be a smooth, compact manifold that admits a nowhere vanishing vector field. Then there exists a smooth map $F: M \to M$ that is homotopic to the identity map and has no fixed point.

42. A **pseudo − Riemannian metric** on a manifold M is a smooth symmetric $2-$tensor field whose value is nondegenerate at each point.

43. Let (M, g) be a connected Riemannian manifold, let G be a Lie group, and let $\theta: G \times M \to M$ be a group action. We say that G **acts by isometries** if for each $g \in G$, the map $\theta_g: M \to M$ is a Riemannian isometry, and G acts **discontinuously** if no $G-$orbit has a limit point in M. If G acts freely, smoothly, and discontinuously on M by isometries, then the quotient map $M \to M/G$ is a smooth covering map.

44. Let E be the total space of the Möbius bundle, which is the quotient of $\mathbb{R}^2$ by the $\mathbb{Z}-$action $n \cdot (x, y) = (x + n, (-1)^n y)$. The **Möbius band** is the subset $M \subset E$ that is the image under the quotient map of the set $\{(x, y) \in \mathbb{R}^2; |y| \leq 1\}$. It is a smooth $2-$manifold with boundary.

45. If M is a compact, smooth, oriented manifold with boundary, then there does not exist a smooth retraction of M onto its boundary.

CHAPTER 11

1. *Suppose* (M, g) *is a Riemannian* n − *manifold. A smooth*
   p − *form* ω *on M is called a* **calibration** *if* ω *is closed and*
   $\omega_q(X_1, \dots, X_p) \leq 1$ *whenever* $(X_1, \dots, X_p)$ *are orthonormal vectors*
   in some tangent space $T_q M$. *An oriented embedded*
   p − *dimensional submanifold* $S \subset M$ *is said to be* **calibrated** *if*
   there is a calibration ω *such that* $\omega|_S$ *is the volume form for the*
   induced Riemannian metric on S.

2. *Let M be a smooth* n − *manifold, and suppose V is a smooth*
   vector field on M such that every integral curve of V is periodic
   with the same period. We define an equivalence relation on M by
   saying that $p \sim q$ *if p and q are in the image of the same integral*
   curve of V. Let $M/\sim$ *be the quotient space, and let* $\pi: M \to M/\sim$
   be the quotient map. Then $M/\sim$ *is a topological* $(n - 1)$ − *manifold*
   and has a unique smooth structure such that π *is a submersion.*

3. *Let M be an* R − *module where R is a PID. We say that M is*
   divisible *if for each nonzero* $a \in R, aM = M$.

4. *Let R be a PID and let M and N be free* R − *modules of the*
   same finite rank. Then an R − *module homomorphism* $f: M \to N$
   is an injection if and only if $N/Im(f)$ *is a torsion* R − *module.*

5. *If R is a commutative ring, then a* **derivation** *on R is a*
   function $\delta: R \to R$ *such that* $\delta(a + b) = \delta(a) + \delta(b)$ *and*
   $\delta(ab) = a\delta(b) + \delta(a)b$.

6. *Let F be an algebraically closed field, and let* $A \in M_n(F)$.
   Then A is nilpotent if and only if all the eigenvalues of A are zero.

7. Let V be a vector space. A linear transformation $E: V \to V$ is called a **projection** if $E^2 = E$.

8. Let M be a finite rank free R − module over a PID R and let $f \in End_R(M)$. Suppose that f is diagonalizable and that $M = N_1 \oplus N_2$ where N_1 and N_2 are f − invariant submodules of M. Then $g = f|_{N_i}$ is diagonalizable for $i = 1, 2$.

9. If R is a ring (not necessarily commutative) and $M \neq \langle 0 \rangle$ is a nonzero R − module, then we say that M is a **simple** or **irreducible** R − module if $\langle 0 \rangle$ and M are the only submodules of M.

10. An Abelian group A is a simple $\mathbb{Z}$ − module if and only if A is a cyclic group of prime order.

11. Let M be an R − module. If M has a composition series let $l(M)$ denote the minimum length of a composition series for M. If M does not have a composition series, let $l(M) = \infty$. $l(M)$ is called the **length** of the R − module M. If $l(M) < \infty$, we say that M has **finite length**.

12. Let M be an R − module of finite length. Then every composition series of M has length $n = l(M)$. Moreover, every chain of submodules of M can be refined to a composition series.

13. Let F be a field. F is said to be **algebraically closed** if every nonconstant polynomial $f(X) \in F[X]$ has a root in F.

14. Let M and N be finite rank R − modules over a PID R and let $f \in Hom_R(M, N)$. If $S \subseteq N$ is a complemented submanifold of N, then $f^{-1}(S)$ is a complemented submanifold of M.

15. Soit A un anneau topologique (non nécessairement séparé).
    On dit qu'un élément x de A est **topologiquement nilpotent** si
    0 est une limite de la suite $(x^n)_{n \geq 0}$.

16. Si A est un anneau, S une partie multiplicative de A, le
    morphisme canonique $Spec(S^{-1}A) \to Spec(A)$ est radiciel.

17. On dit qu'un anneau topologique A est **linéairement
    topologisé** s'il existe un système fundamental de voisinage de
    0 dans A formé d'idéaux (nécessairement ouvert).

18. If **g** is any finite − dimensional Lie algebra, then the
    disconnected Lie groups whose Lie algebras are isomorphic to
    g are precisely the extensions of the connected ones by
    discrete groups.

19. We define a **foliation** of dimension k on an n − manifold
    M to be a collection of disjoint, connected, immersed
    k − dimensional submanifolds of M (called the **leaves** of the
    foliation) whose union is M and such that in a neighborhood of
    each point $p \in M$ there is a smooth chart (U, φ) with the property
    that $\varphi(U)$ is a product of connected open sets
    $U' \times U'' \subset \mathbb{R}^k \times \mathbb{R}^{n-k}$, and each leaf of the foliation intersects
    U in either the empty set or a countable union of
    k − dimensional slices of the form $x^{k+1} = c^{k+1}, ..., x^n = c^n$.
    (Such a chart is called a **flat chart** for the foliation.)

20. If M and N are connected smooth manifolds, the collection
    of subsets of the form $\{q\} \times N$ as q ranges over M forms a
    foliation of $M \times N$, each of whose leaves is diffeomorphic to N.

21. If G and H are Lie groups, and there exists a surjective
    Lie group homomorphism from G to H with kernel G_0, we say that
    G is an **extension of** G_0 **by** H.

22. **Herman – Yoccoz Theorem**
    If f is a real analytic diffeomorphism of $\mathbb{R}/\mathbb{Z}$ and if the
    rotation number ρ is Diophantine, then f is real analytically
    conjugate to the rotation $t \longmapsto t + \rho \pmod 1$.

23. Let V be a complex linear space. Then by a **sesquilinear**
    form on V we mean a map $\beta: V \times V \to \mathbb{C}$ which is linear in the
    first variable, and conjugate linear in the second, i.e.,
    $\beta(v, \lambda w + w') = \bar{\lambda}\beta(v, w) + \beta(v, w')$ for all $v, w, w' \in V, \lambda \in \mathbb{C}$.

24. Let $\varphi: G \to H$ be an immersive homomorphism of Lie groups.
    Then φ is a submersion if and only if $\varphi(G)$ is an open subgroup
    of H.

25. A **fundamental chain** is an infinite sequence $A_1, A_2, \ldots$ of
    disjoint transverse arcs with the property that the corresponding
    neighborhoods are nested: $N(A_1) \supset N(A_2) \supset N(A_3) \supset \cdots$, and with
    the property that the diameter of A_i tends to zero as $i \to \infty$.

26. Pour qu'un morphisme $f: X \to Y$ soit séparé, il faut que pour
    tout ouvert U sur lequel Y induit un préschéma séparé, le
    préschéma induit $f^{-1}(U)$ soit séparé, et il suffit qu'il en soit ainsi
    pour tout ouvert affine $U \subset Y$.

27. On dit qu'un préschéma est **artinien** s'il est affine et si son
    anneau est artinien.

28. Soit X un K − préschéma algebrique. Pour qu'un point $x \in X$
    soit fermé, il faut et il suffit que K(x) soit une extension
    algebrique de K, de degré fini.

29. Étant donné un corps K, on appelle K − **préschéma**
    algebrique un préschéma X de type fini sur K; K est appelé le

corps de base de X. Si en outre X est un schéma (ou, ce qui revient au même, si X est un K − schéma), on dit aussi que X est un K − **schéma algébrique**.

30. Soit X un préschéma localement noethérien. toute $\mathcal{O}_X$ − algèbre quasi − cohérente de type fini $\mathcal{B}$ est alors un faisceau cohérent d'anneaux.

31. Soit X un espace de Kolmogoroff noethérien. Pour qu'un point $x \in X$ soit isolé, il faut et il suffit que $\dim_x X = 0$.

www.ingramcontent.com/pod-product-compliance
Lightning Source LLC
Chambersburg PA
CBHW070920180526
45168CB00005B/2079